AF262814

union
square
kids

NEW YORK

Union Square Kids
Hachette Book Group
1290 Avenue of the Americas, New York, NY 10104
unionsquareandco.com
@unionsqandco

First Edition: June 2026

Union Square Kids is an imprint of Grand Central Publishing, a division of Hachette Book Group, Inc.
The Union Square Kids name and logo are registered trademarks of Hachette Book Group, Inc.

The publisher is not responsible for websites (or their content) that are not owned by the publisher.

Union Square Kids books may be purchased in bulk for business, educational, or promotional use.
For information, please contact your local bookseller or the Hachette Book Group Special Markets Department
at special.markets@hbgusa.com.

Text by Ruth Martin
Book design by Clarisse Hassan
Edited by Rosie Neave

Library of Congress Cataloging-in-Publication Data has been applied for.

ISBN 978-1-4549-6173-4

Printed in Guangdong, China
TLF 02/26

2 4 6 8 10 9 7 5 3 1

For my daughters, Tabitha and Miretta, and the next generation of strong young women—R.M.

5 Minute Genius Stories

ADA LOVELACE

Written by Ruth Martin

Illustrated by Sam Rudd

How to use this book

In this book you'll find **ten genius stories** to read, each one just **5 minutes long**.

At the end of each story, explore an informative **"all about"** spread.

Want to learn more? Turn to the back of the book to discover a **timeline of key events**.

union
square
kids

NEW YORK

WHO WAS ADA LOVELACE?

Ada Lovelace combined skilled scientific thinking with creative imagination to develop the first computer program. She described the potential for computers long before the technology itself existed, and inspired the computer industry we know today.

Ada combines her mother's intelligence and her father's imagination.

Page 8

Ada became fascinated by flight and writes a book called *Flyology*.

Page 18

Ada is bedridden but uses the time to learn all about space science.

Page 28

Ada tours the textile factories and is fascinated by the programmable looms.

Page 38

Ada meets Charles Babbage and learns about the Difference Engine.

Page 48

Ada begins translating an article about the Analytical Engine.

Page 58

Ada works hard on her own notes about the Analytical Engine.

Page 68

Ada begins to question whether machines could think for themselves.

Page 78

Ada attends the Great Exhibition and realizes the power of her own mind.

Page 88

Ada's revolutionary ideas live on in the computerized world.

Page 98

WHICH 5-MINUTE GENIUS STORY WILL YOU READ TODAY?

A CURIOUS CHILD
Creativity and Logic Combine

There was great excitement in England when a little girl named Augusta Ada Byron was born in London in 1815.

Known by her middle name of Ada, she was the daughter of famous poet George Gordon Byron and an upper-class lady named Anne Isabella Milbanke. The two were as different as can be.

Known as Lord Byron, Ada's father wrote poetry about love and adventure.

He was rebellious and dramatic—a bit like a rock star!

Known as Isabella, Ada's mother loved math and science rather than poetry. Lord Byron called her his "Princess of Parallelograms."

Ada's mother and father separated when she was still a tiny baby. Her father left for Italy, and Ada stayed with her mother.

As Ada grew up, her mother worried that Ada would inherit her father's passionate creativity and that this would lead to rebellious behavior.

Would Ada grow up to be logical, like her mother, or creative, like her father?

Ada's mother felt strongly that girls should be raised to be intelligent, and develop logical, sensible minds.

She had a particular interest in astronomy, and began to teach Ada about the science of space.

Ada looked at the moon and stars. She was fascinated by space science, but her imagination was also taking flight.

Ada's mother insisted that she study with private tutors—experts who could bring their subjects to life.

She was taught arithmetic, grammar, spelling, reading, music, geography, drawing, French, and music.

Ada was extremely capable at her lessons—and she also had a very active imagination.

Ada felt bored by her lessons and found it hard not to fidget. She desperately wanted to get outside and explore—she couldn't sit still any longer!

When she was released from her lessons, Ada would often play outside with her beautiful fluffy cat, Mrs. Puff.

Ada had a feeling there was far more to learn outside in the world than in any textbook.

When Ada was eight years old, news came that Lord Byron had died. Though she never really got to know her father, Ada shared some characteristics with him.

Like him, Ada was very curious about what she later described as the "unseen worlds around us." When Ada saw a rainbow, she was immediately curious about how it appeared in the sky.

While most people just enjoyed such a sight, Ada could not rest until she understood all about it.

Her natural curiosity combined her mother's intelligence and her father's imagination.

Ada didn't have to choose between being creative and being logical—she could be both.

Education for ALL

In the early 1800s, most children worked to support their families. They did not attend school, and often could not read or write.

Schools were mostly for the sons of **wealthy families.**

Boys were taught classical subjects such as Latin and Greek, as well as history, science, and math.

Some upper-class **girls** would be taught at
home by a governess. These lessons usually
involved memorizing basic facts from books.

They were also
expected to learn
to play **music** and to
sew, and to develop
good **manners**.

Compared to other upper-class
women of the time, Ada's
mother was very **unusual**.

She was extremely well
educated, and she believed all
children should be encouraged
to **think** rather than simply
memorizing information.

A YOUNG MIND TAKES FLIGHT

Looking Up for Inspiration

In 1826, when Ada was 11 years old, she set off with her mother for a Grand Tour of Europe.

Ada's mother wanted Ada to see the world for herself, to learn about different cultures, languages, and people.

As they traveled by boat, train, and carriage, Ada's gaze was drawn up to the sky. There she saw birds soaring past them—brightly colored and of all shapes and sizes.

The birds' flight was seemingly effortless. Ada wondered: how do they do that?

She became fascinated by flight and began to sketch ideas in her notebook.

Back at home, Ada was
full of ideas bursting
to come to life.

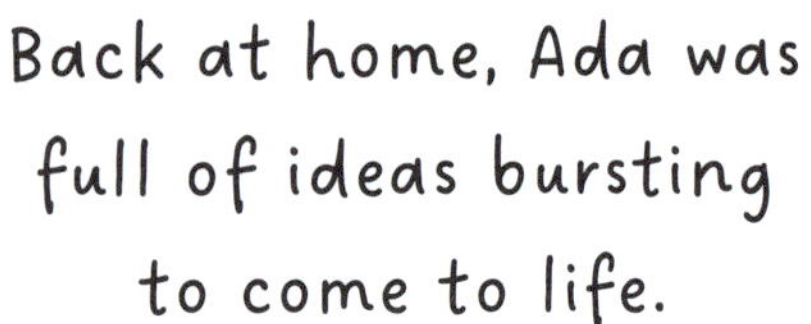

She asked her tutors lots
of questions about flight and listened
keenly to all they could tell her.

She studied bird
anatomy, from their body
shapes and wingspans, to
what their feathers did.

Then Ada had to develop her
math skills to understand how
a bird's wing-to-body ratio
helped it to take flight.

Ada even began to imagine herself flying!
She decided to design and build her own wings and
experimented using paper, silk, feathers, and metal wire.

Ada grew up at a time of great industrial progress. She saw steam engines powering new machinery and railways transporting people faster than ever before.

Ada took inspiration from this "mechanical age." Ideas tumbled through her mind, moving as quickly as the wheels, cogs, and engines she saw all around her.

Ada put all her ideas together to make a book, which she proudly named Flyology.

This ambitious creation contained Ada's observations and ideas about how steam power could create machinery that could fly, just like a bird.

Ada explained everything carefully, and even illustrated the book herself.

Ada's book contained a particularly exciting idea—
something that had never been invented before—
a flying mechanical horse!

Ada buzzed with excitement,
imagining a world where
people could fly around on
machines of her creation.

Ada proudly showed
her mother her
exciting ideas.

Lady Byron was pleased with all the math and science that her daughter had learned.

But she was concerned that Ada's thoughts were running away with her.

She didn't realize that these flights of fancy were part of what made Ada's mind special.

Ada's design imagined the future of air travel almost eighty years before the first plane would take flight.

The Wonder of FLIGHT

People have always been fascinated with the idea of soaring into the sky. Only in the Victorian age did it begin to seem possible.

Ada's studies would have taught her that birds use their wings to generate two forces, lift and thrust.

These forces must overcome two counterforces—**gravity** and **drag**—to allow for flight.

In flapping flight, birds use their wings like oars to "row" through the air. Most birds alternate between **flapping** and **gliding**, to save energy.

Some early attempts at human flight were disastrous. As early as the 1600s, people were making wings and trying to fly, with many dying in their attempts.

A hot air balloon flies because the air inside it is heated, making it lighter than the surrounding air.

The first successful hot-air balloon flight was in 1783. The craft was designed by French brothers Joseph-Michel and Jacques-Étienne Montgolfier.

In Germany, Otto Lilienthal made successful glider flights starting from 1891.

The Wright brothers, designers of the first airplane, credited Lilienthal as one of their inspirations.

THINKING IN CIRCLES
Finding Space to Imagine

In 1829, when Ada was fourteen, she became very ill. Her head ached and her eyesight was blurry.

Doctors struggled to find a cure for Ada's illness, so she was advised to stay in bed and rest.

Some days, she couldn't even see to read or move her legs to walk.

For over a year, Ada found herself stuck at home in very poor health.

With nowhere to go, and very little company,
Ada decided to use her time to learn.

She loved solving scientific problems
and soaking up all she could read
about the wider world.

She even taught
herself languages,
including Latin
and German.

It took three years
for Ada to begin
to feel well again. By
then, her mind was
set on the future and
hungry for greater knowledge.

Ada's mother traveled
a lot, leaving Ada in
the care of her friends
and tutors.

Ada nicknamed
three of these friends
"the Furies," after the
goddesses of the underworld
in Greek mythology.

She thought the Furies were controlling
and unkind, forcing her to behave
against her true nature.

The Furies were under
strict instructions
from Ada's mother.

They would punish Ada
to stop her fidgeting and
report to her mother if
Ada's imagination began
to run wild.

Ada escaped into her studies—especially when her instructor was one very unusual woman. Mary Somerville was a famous Scottish scientist and mathematician.

Ada listened with amazement as Mary described the Solar System, with its many different planets moving in circles around the Sun.

Another tutor, Augustus De Morgan—famous for his work
in logic and algebra—took time to make sure there were
no gaps in Ada's understanding of math.

She could see the
mysteries of space being
brought into logical
mathematical order.

$$A - (B \cup C) = (A-B) \cap (A-C)$$
$$A - (B \cap C) = (A-B) \cup (A-C)$$

Ada was fascinated by
the possibilities of bringing an
imaginative mindset to math and science.

When Ada turned 17, she attended a grand royal event where she was presented to King William IV and Queen Adelaide.

It marked the start of Ada's adulthood, meaning she would soon be expected to marry and have children.

Ada knew that taking on these responsibilities would tear her away from a world of exciting inventions and inspiring ideas.

Lady Byron advised Ada to study only science and never poetry. But Ada wanted to be free to think creatively as well as logically.

Ada wanted space to be herself and challenge the rules of science, society, or anything else that stood in her way.

Our place IN SPACE

When scientists were making exciting discoveries about our Universe in the 1800s, planetariums (mechanical models of the Solar System) became popular.

A **planetarium** featured each planet in orbit around a central Sun.

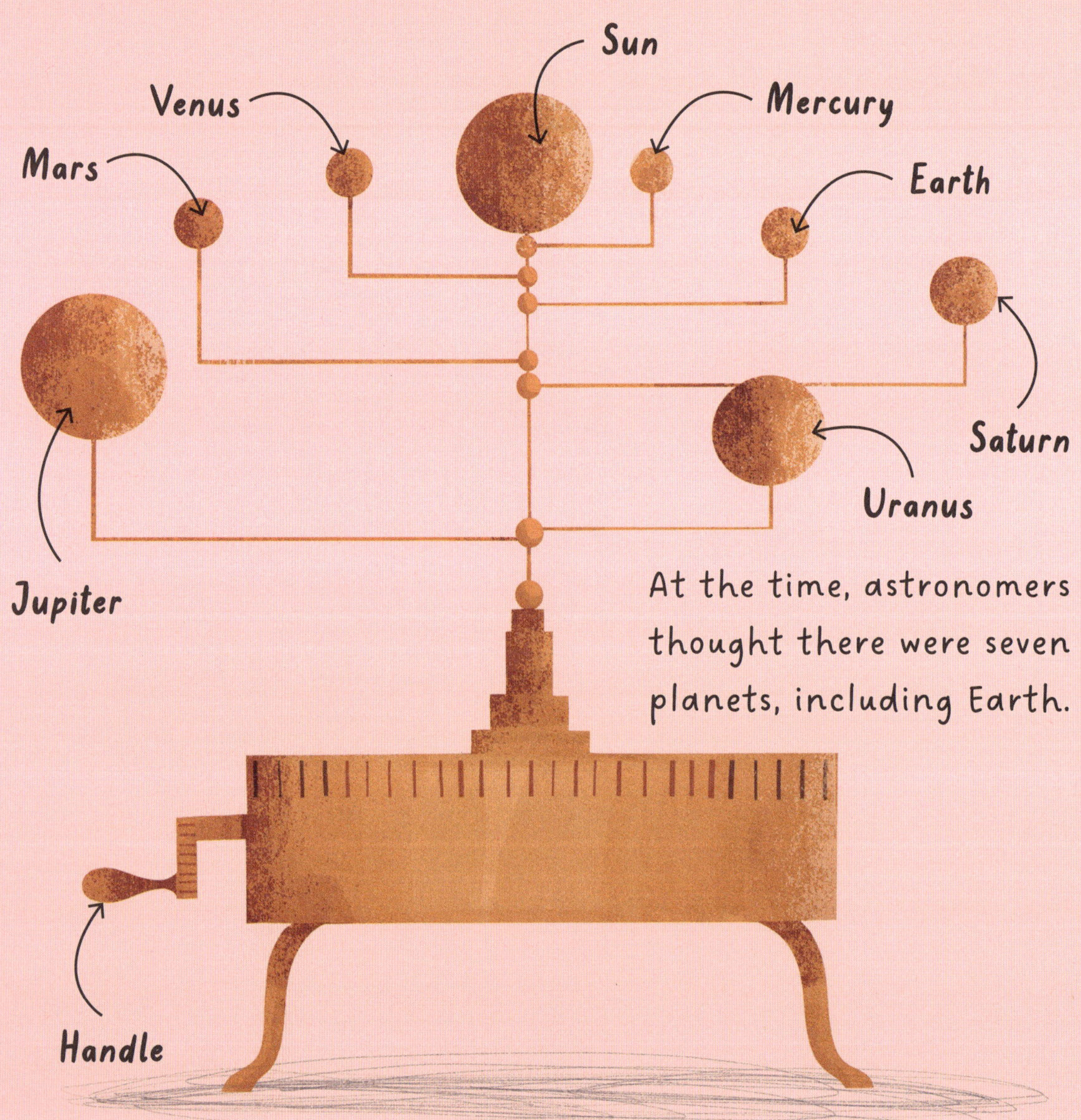

At the time, astronomers thought there were seven planets, including Earth.

By turning a handle, the mechanism allowed the planets to rotate to demonstrate their **motion in space**.

Ada's friend and tutor Mary Somerville (1780–1872)
predicted the existence of an **eighth planet**.

Mary described how difficulties in calculating the
position of **Uranus** might suggest the presence of
an as-yet undiscovered planet.

Her ideas were later proved correct when
Neptune was discovered in 1846.

PUNCHING PATTERNS
Programming the Future

Ada's mother rushed to tell Ada to pack a bag—
they were leaving for a tour of the north!

The year was 1834,
and Ada was 19 years old. Northern
textile factories buzzed with activity
as towns all over the country were
becoming more industrialized.

Ada's mother was fascinated
by these social changes and
wished to see them for herself.

The hustle and bustle of the noisy
station was exciting to Ada, and
her eyes darted around to see the
trains coming and going.

The city was smoky and
dirty. Ada noticed the hem
of her skirt turning black as
she climbed into the train.

From the carriage window, Ada saw
rows and rows of chimney stacks,
busily puffing smoke into the sky.

Lines of people were filing into the factories to start work.

On their tour of the textile factories,
Ada and her mother saw huge
machinery clanking and
whirring amid the
busy workers.

Some workers were unhappy at being replaced by technology that could do the work of many people.

One master weaver had even thrown his shoe into a loom as a protest, breaking all the threads.

Ada was very interested by the idea of a machine that could perform a complicated task with such efficiency.

One particularly special machine was called a Jacquard loom.

It produced
richly colored
cloth with
a detailed
floral design.

But how could
a machine be
taught to create
such complex
patterns?

It sparked a flurry of questions . . .

The key was a card punched with a series of holes
that slotted into the loom. From this, the machine knew exactly
what to do to weave the elaborate patterned fabric.

Ada returned home, her mind whirring with
everything she had seen on her trip . . .

. . . from the engines
at the railway station,
allowing people to
travel at speed all
over the country . . .

. . . to the machines
that were making
manufacturing faster
and more efficient.

And she was fascinated by the punch
cards of the Jacquard loom, which
could instruct the machines to
produce such luxurious textiles.

Ada realized that the
programmable nature of this new
technology could change the world.

Textile TECHNOLOGY

Invented in 1804 by French weaver and merchant Joseph-Marie Jacquard, the Jacquard loom revolutionized the textile industry.

Joseph-Marie Jacquard

Warp and **weft** refer to the direction
of the woven threads in a fabric.

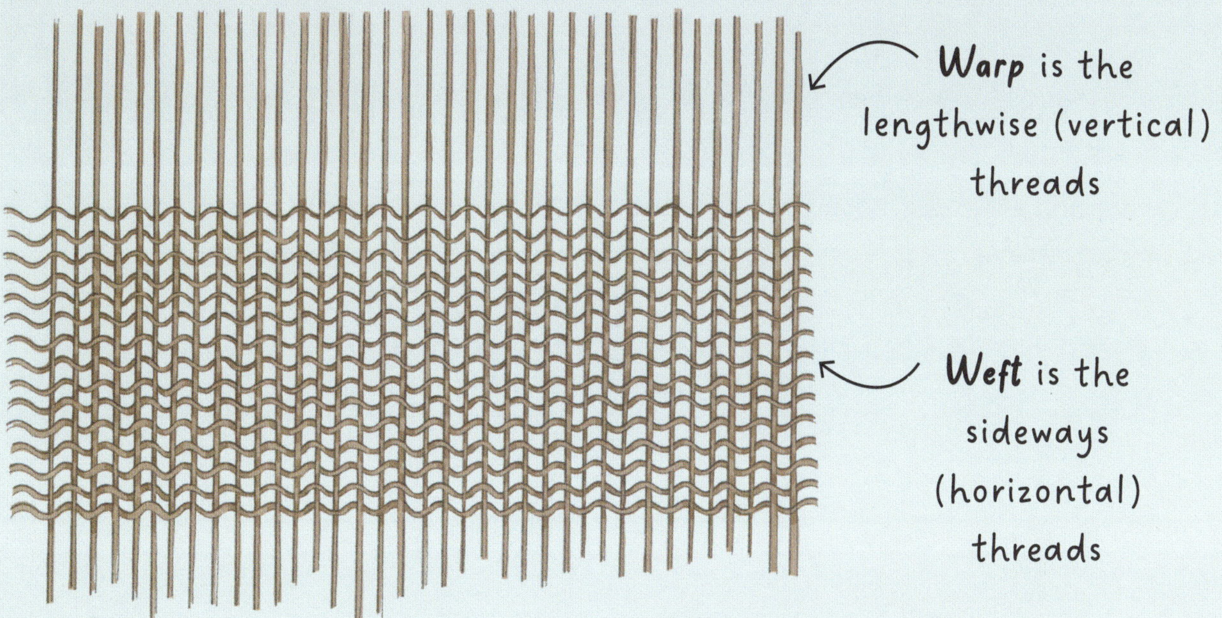

Warp is the
lengthwise (vertical)
threads

Weft is the
sideways
(horizontal)
threads

Punch cards placed inside
a loom set the position
of the warp threads to
create patterns in the
woven cloth.

Each space on the pattern
cards had either a hole, or
no hole, forming what is
called a **binary system**.

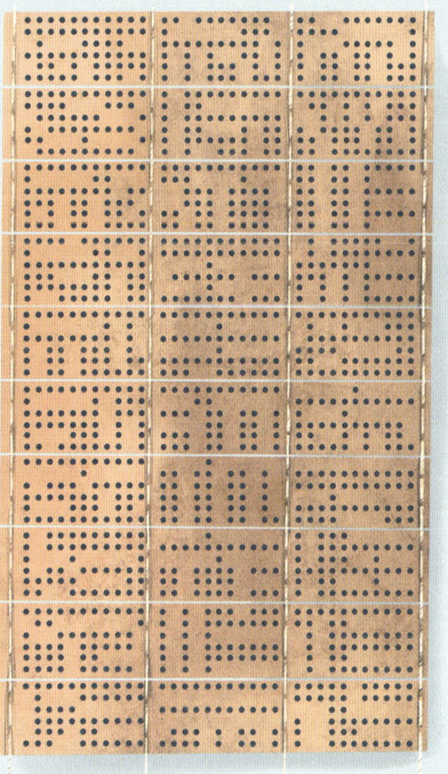

The Jacquard loom inspired early designs
for computers, which are also **programmed**
to carry out automated processes.

A MEETING OF MINDS

The First Calculating Machine

In 1833, when Ada was 17, she and her mother were invited to a party in London by the famous mathematician and inventor Charles Babbage.

Charles was a "polymath" (an expert in many subjects). Like Ada, he had an active mind, buzzing with ideas.

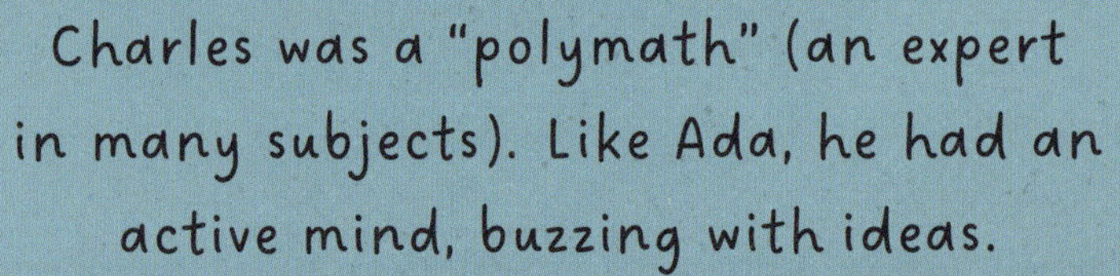

When Ada arrived, the house was full of highly educated guests. She found their conversations very exciting.

Everybody was talking about Charles's latest invention. The Difference Engine was a machine that could perform complex calculations with numbers.

In a time with no calculators, most people couldn't imagine how a machine could do math.

Charles led Ada to his study to see his new invention.
It was an intricate piece of machinery, unlike anything
Ada had seen before. "What can it do?" she asked.

Charles explained that
mathematical calculations
took so long and were so
prone to mistakes, that
he invented his Difference
Engine to help.

The machine could process the calculation of
numbers, and not just single numbers but series of
numbers to solve complex problems.

Charles explained that
this was just a model—
the real machine would
be much bigger and very
expensive to build.

Ada imagined many uses for such an invention,
and she asked Charles more and more questions.

Ada was soon invited back to visit Charles. They loved to talk in great detail about his invention, solving problems and wondering how to build a real working model.

Charles was excited to tell Ada that he had received a large amount of money from the government to build his machine.

While Charles worked on his machine, Ada
imagined the future of what it could do.

Beyond just helping with
navigation and industry, Ada
felt the Difference Engine had
all sorts of exciting potential.

Charles and Ada exchanged many letters
discussing their ideas. Messengers dashed
between their houses with the mail.

Charles could be grumpy and rude. He upset
his engineer and even the prime minister!

Ada had a more
positive approach,
determined to
move forward with
the designs.

Charles called Ada the
"Enchantress of Number."
Ada signed off letters to
Charles, "your puzzle-mate."
Together, they were
a dream team.

From their
meeting at a party,
Ada and Charles grew
a friendship that led
to some very big ideas.

The Difference Engine had set the cogs in
motion for the future of computer design.

The DIFFERENCE ENGINE

Charles and Ada never managed to build a complete model of the Difference Engine in their lifetimes.

The small **prototype** from Charles's home was given to the Science Museum in London, U.K., after Charles's death.

Years later, the Science Museum followed his plans to build a **full-scale model**.

Completed in 2002, the result was a vast machine with **thousands** of parts.

Numbers were represented
by metal wheels arranged
in **columns**.

The machine was made
of **iron**, **steel**, and **bronze**.

It was operated by
a large **handle**.

A built-in **printer** output the
results of each calculation.

Completed, the machine
was 11 feet long and
weighed nearly 5 tons.

Most importantly, the model worked! It could
solve **equations** and print the results in the
form of mathematical tables. Charles and
Ada would have been thrilled.

THE ANALYTICAL ENGINE
An Invention Needs a Voice

In 1835, when Ada was 19, she and her mother attended a ball at the home of a wealthy textile businessman.

Ada's mother knew successful gentlemen would be there. Ada was now a young woman, and society expected her to make a good marriage.

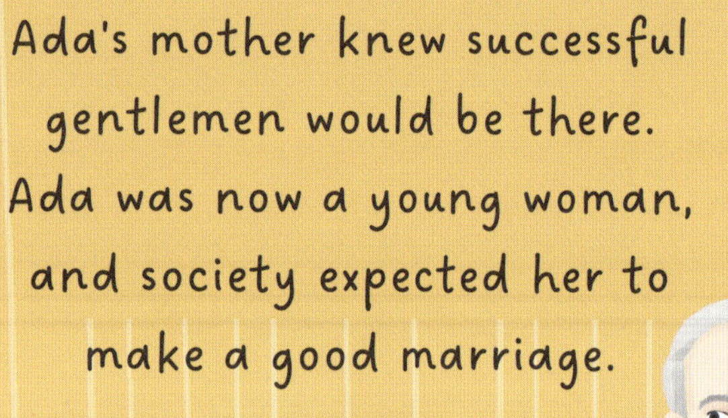

Ada wore a beautiful ball gown and was very popular with the party guests. Soon she was introduced to William, known as Lord King, a handsome man from a wealthy and respectable family.

William asked Ada to dance, and soon they were chatting and laughing. Ada's mother was pleased with the match. She felt William would be an excellent husband for Ada.

William proposed to Ada soon after they met and they married that same year.

In 1836, the year after their marriage, Ada and
William had their first child. The baby boy was
named Byron, after his famous grandfather.

By 1839, Ada and William had two more children: a girl
named Anne Isabella and another boy, Ralph Gordon.
Ada found herself very busy with three young children.

William had been granted a new title—Earl of Lovelace—by the Prime Minister, so Ada became the Countess of Lovelace. These titles symbolized that their family had power and importance in society.

But despite their success, Ada missed her old freedom and her studies.

She longed for her mind to be set alight again . . .

. . . and luckily that was just about to happen.

Ada's friend Charles Babbage was working on an exciting new invention.

It was still at the planning stage, and was called the Analytical Engine.

The Analytical Engine was a steam-powered machine that used punch cards, like those used by a Jacquard loom to produce patterns on cloth. The Engine's cards programmed calculations.

When Charles described the machine
to Ada, he used words just like those
used to describe a textile factory.

The two main parts
of the machine were
known as the "mill," where
calculations took place, and
the "store," where numbers
could be held or recorded.

Charles knew he had
planned a very useful
machine, but he was
not sure how to bring
his designs to life.

Ada could see it was possible . . . and she
began to imagine how important this
Analytical Engine would be.

Charles felt frustrated that enthusiasm and funding for his inventions seemed to have run out of steam in England.

So he traveled to Italy, where he lectured to scientists who were more excited by his ideas.

Inspired by Charles's lecture, an Italian mathematician named Captain Luigi Menabrea wrote an enthusiastic article to help other scientists understand the machine's great potential.

But the article was in French, so it wouldn't help Charles in England. Charles soon realized he needed Ada's help to translate it.

Ada was captivated by the project.

With all her skills in math, science, and languages, she knew she could put a spotlight on Charles's invention and make the world pay attention!

Ada was uniquely qualified to explain the significance of the Analytical Engine to the world.

What is a COMPUTER?

Charles Babbage's Analytical Engine was never fully built—but it was the first concept of a computer.

Before Charles's idea, there were no machines to do math. The word "computer" referred to a **person** who performed calculations.

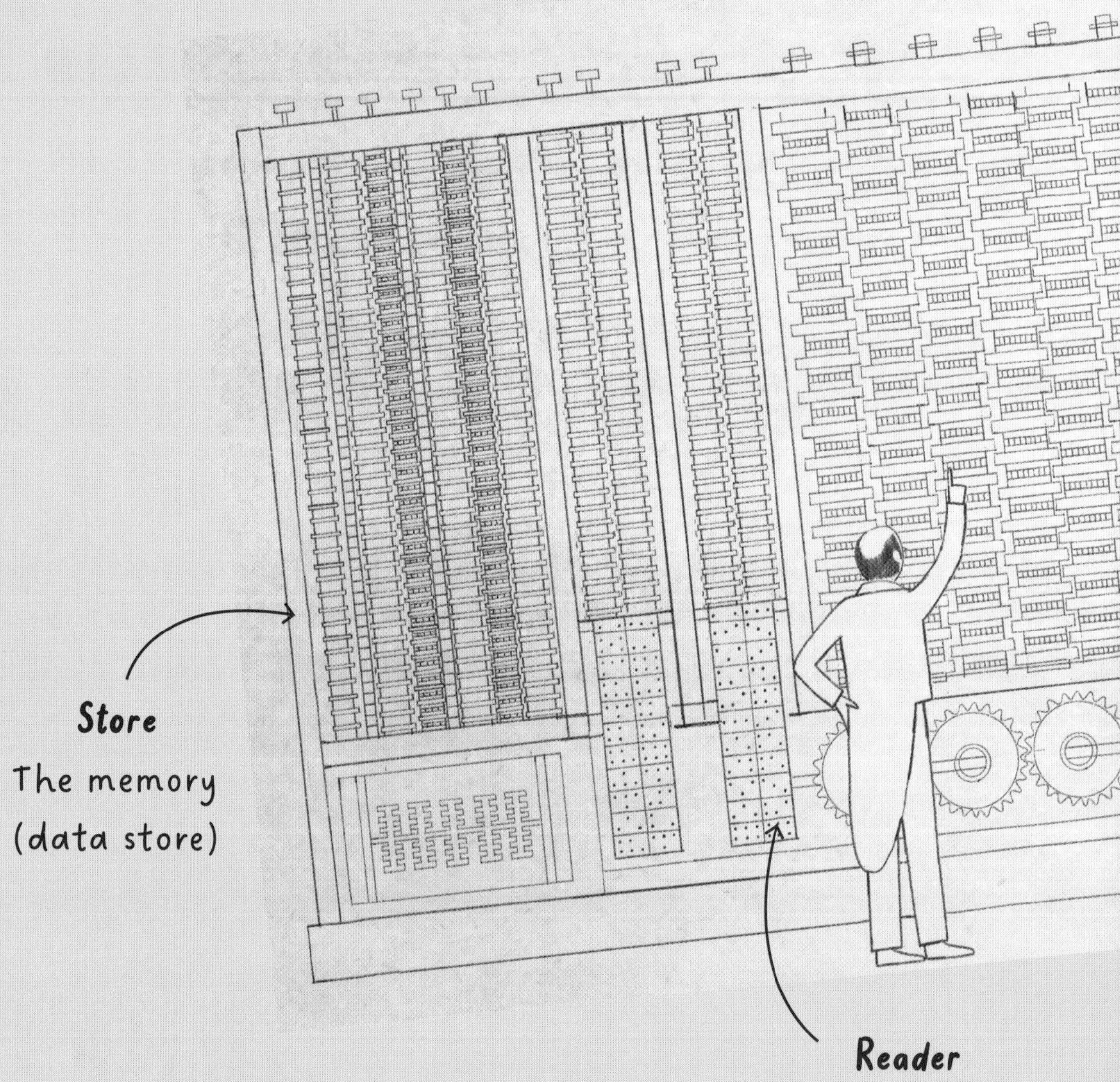

Charles's plans showed how gears, levers, and steam power could be used to complete calculations very **accurately**.

The vast machine would have been more
than 20 feet long and over 10 feet high.
Since Charles's early ideas, computers have
developed into much smaller electronic devices
that can process **data** at great speed.

Mill
The calculating
unit

Printer
The output for
a record of the
calculations

Modern computers help
us in everything from
simple personal tasks
to ground-breaking
missions in space.

GENIUS NOTES
From Translator to Teacher

Ada began translating Captain Luigi Menabrea's French article about her friend Charles Babbage's Analytical Engine into English in 1842, at 26 years old.

Ada was excited.
At the time, women were
rarely given the opportunity
to work on scientific papers,
and she knew she was the
perfect person for the work.

Ada was highly skilled in math, science, and languages. Her writing was clear and engaging—ideal for explaining the complex ideas behind the invention.

As Ada began work on the translation, she could see that the French article was far too complicated for most people.

Ada knew that if her translation could explain the Analytical Engine successfully, it might persuade people to support Charles in bringing the designs to life. So she began adding her own explanations.

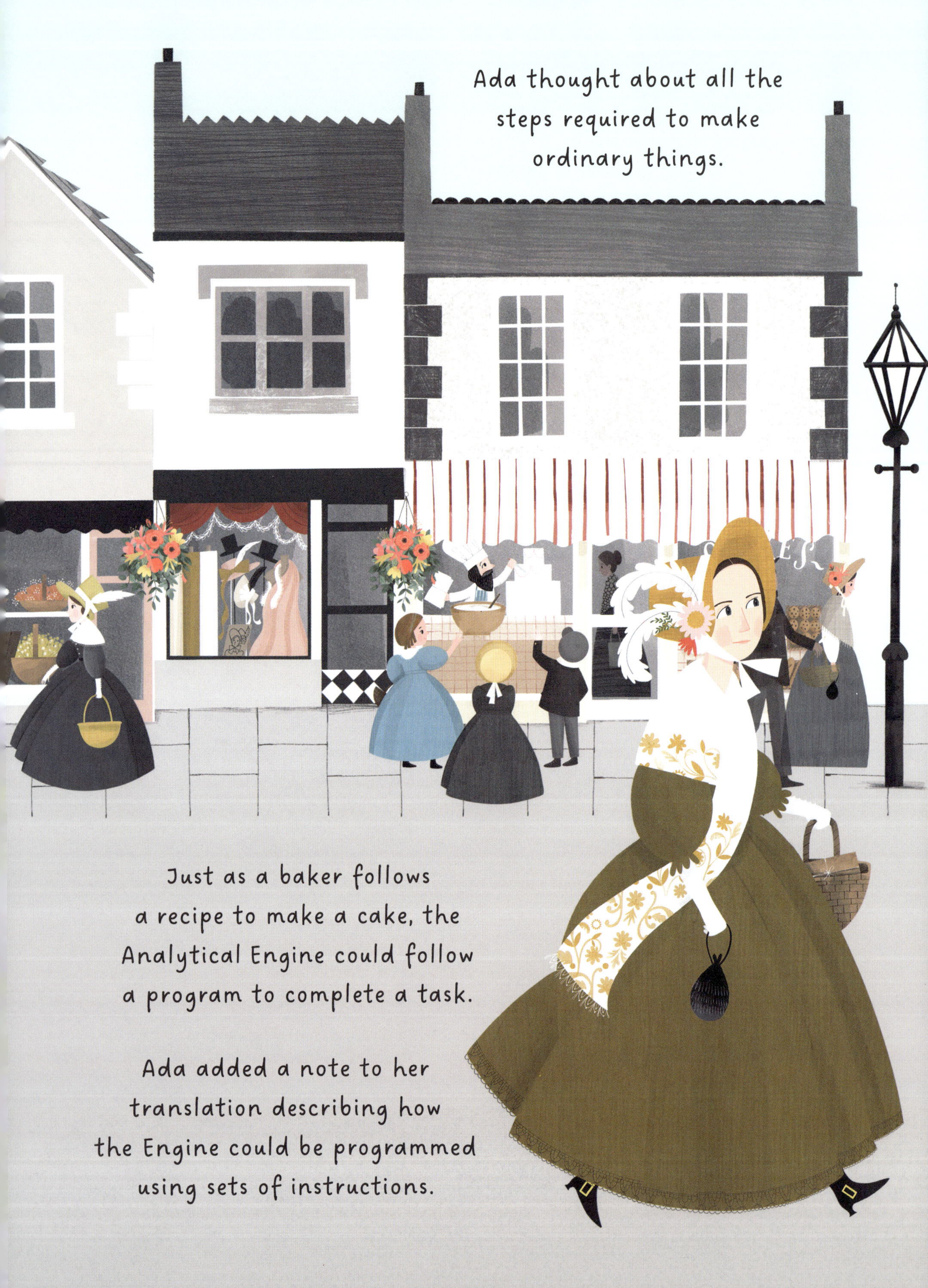

Ada thought about all the
steps required to make
ordinary things.

Just as a baker follows
a recipe to make a cake, the
Analytical Engine could follow
a program to complete a task.

Ada added a note to her
translation describing how
the Engine could be programmed
using sets of instructions.

She gave an example of step-by-step methods—
which today might be likened to a recipe—that could
be used to calculate a sequence of figures.

By giving the machine
different sets of
instructions, it could
be used for many
different purposes.

Although Ada couldn't have known the significance of her
discovery at the time, her notes had explained the very first
detailed computer program in the world.

Ada drew inspiration from the Jacquard loom
she saw on a childhood trip at the textile
mills with her mother. The loom could
be given instructions to create many
different woven fabrics.

Ada's notes explained
how Charles's new invention
would be a step beyond
his previous invention, the
Difference Engine. This
one would be capable of
performing far more than one
fixed calculation at a time.

She wrote, "The Analytical
Engine weaves algebraic
patterns, just as the Jacquard
loom weaves flowers and leaves."

Ada described how the machine could
potentially work with letters or symbols, making
it far more than a mathematical machine.

One day, she thought, this machine could be used to make other sorts of art, such as music.

As she looked over her work, Ada noticed her notes were longer than the article itself!

She was no longer a student—she had become a teacher herself, passing on her complex scientific knowledge.

One windy day, as the leaves danced around her feet,
Ada visited Charles Babbage. In her basket, she carried
her latest notes for her translation.

She regularly made this
walk to see Charles and ask
him questions as she worked
on her article.

Ada's thoughts whirled like
the rushing leaves—she had
so many questions that
Charles would never be able
to answer them all.

Ada's notes and ideas had begun
to take on a life of their own, beyond
Captain Menabrea's original article.

Back at her desk that
evening, Ada put her notes in
order and added her initials
to each one—A.A.L, for
Augusta Ada Lovelace.

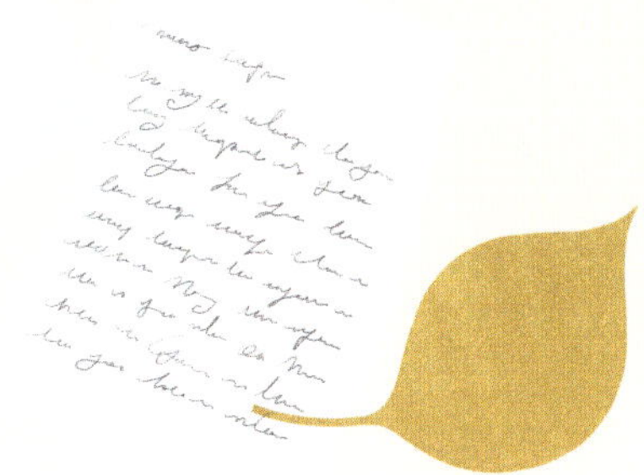

She wanted
everyone to know
which parts were
her own work.

Ada was starting to take
ownership of her ideas.

Computer PROGRAMS

Ada described computer programs and algorithms before computers existed.

A **computer program** is a list of instructions written as a code that tells a computer what to do. The Analytical Engine could be **programmed** using punched cards.

An **algorithm** is a step-by-step process to complete a task or solve a problem.

Just as a **recipe** allows you to make a cake, a computer follows instructions from a **program** with algorithms to complete different tasks.

Ada's important notes were labelled from **A to G**.

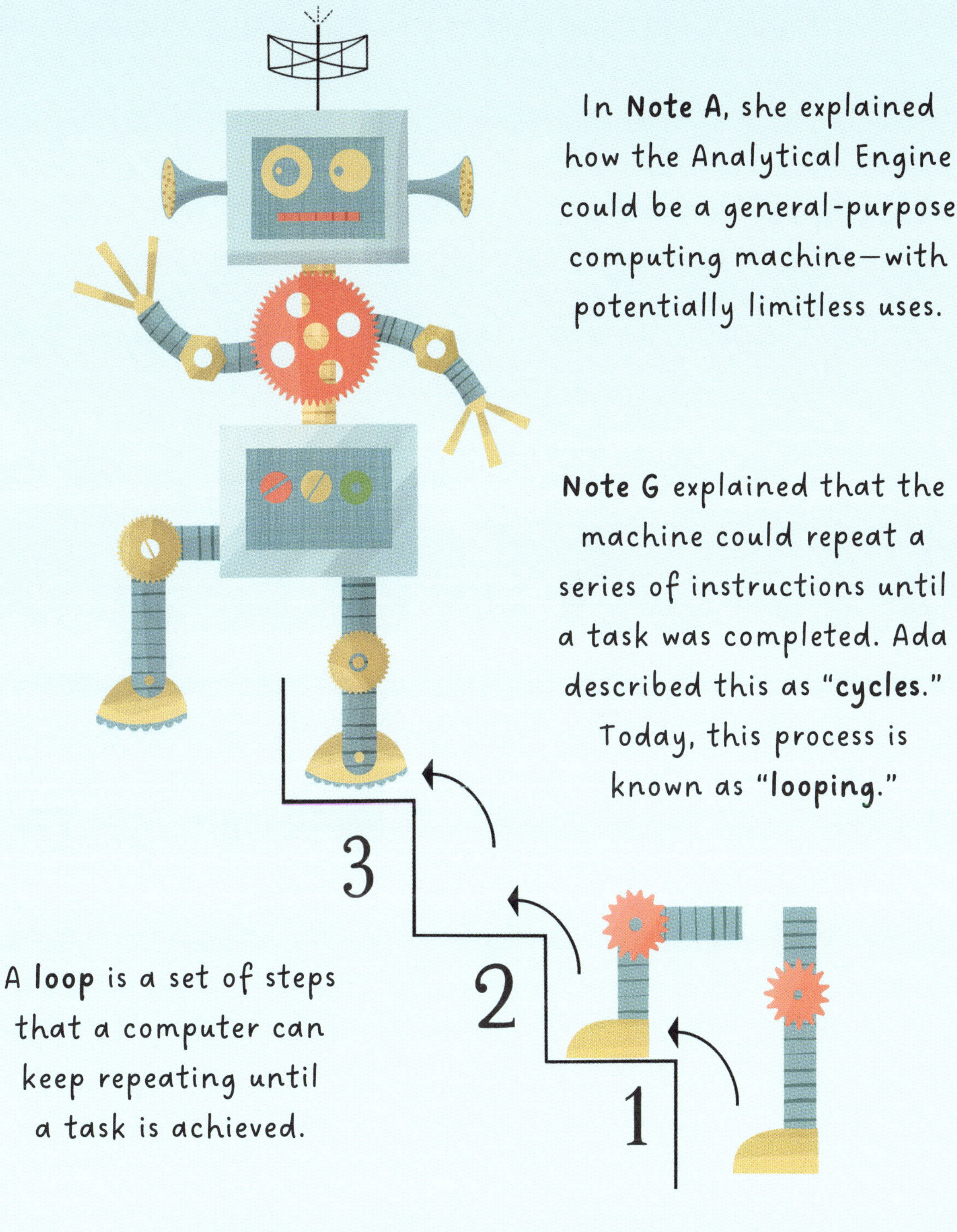

In **Note A**, she explained how the Analytical Engine could be a general-purpose computing machine—with potentially limitless uses.

Note G explained that the machine could repeat a series of instructions until a task was completed. Ada described this as **"cycles."** Today, this process is known as **"looping."**

A **loop** is a set of steps that a computer can keep repeating until a task is achieved.

For example, to make a robot go upstairs, it must first be programmed to go up one step. A loop allows that instruction to be **repeated** until the robot reaches the top.

A NEW WAY OF THINKING

Imagining the Future

Ada's childhood imagination had been rich and free. The flying mechanical horse she had imagined in the book she wrote as a child, *Flyology*, was an early sign of how she could pair creativity with technical understanding.

Now, in 1843, her work was tied up in facts and figures as she translated the scientific article about her friend Charles Babbage's invention.

Charles's invention—the Analytical Engine—had required great feats of engineering know-how and mathematical skill.

It was Ada's imagination that took the next leap in visualizing how his machine could be programmed for many different uses.

As a team, Charles and Ada combined his
engineering genius with her creative imagination.

Ada brought a magical touch to the
work, with her open-minded approach
and visionary ideas.

Charles called Ada his "lady fairy,"
and she even signed off her letters
to him with this magical nickname.

Ada knew the Analytical Engine needed
a human operator—someone to do the
thinking and set the machine working.
It could not create by itself.

But this started her
thinking . . . what is
the difference between
a machine and a
human mind?

Although Ada felt the machine could not have a mind
of its own, the questions she was asking were the start
of a radical idea for other scientists to consider.

For the Analytical Engine to become a reality, Charles had to complete the design. Then it would need an engineer's hands and expertise to bring it to life. But what might happen next?

Ada wrote in a letter to Charles: "I want to put in something about Bernoulli's Numbers . . ." These were a complex sequence of figures named after Swiss mathematician Jacob Bernoulli.

Ada wondered—since the engine was capable of calculating the Bernoulli numbers by itself, what else could it learn to do? Could such a machine begin to "think" for itself?

After nine months of work, Ada felt her translation was complete and ready to be published.

One day in August 1843, Ada heard a knock at
the door and found a parcel had been delivered.
She and her husband William opened it together.

The parcel contained printed
copies of her translation,
published in a book called
Scientific Memoirs.

It was rare for a woman to
have scientific work published,
so Ada's work would be notable
among the scientific community.

Of the sixty-six pages of the article, forty-one were by Ada, marked with her initials.

Ada and William marveled at seeing her exciting ideas in print. They shared copies with friends and sent one to Ada's mother.

Ada's pioneering ideas had been noticed, and now scientists could start to build on her ideas.

Artificial INTELLIGENCE

Ada's notes, and the responses to her work by other scientists, were the stepping stones to the invention of Artificial Intelligence (A.I.).

Humans learn from **experience**, but computers cannot do this.

Using an A.I. **program**, a computer "learns" (self-improves) by finding **patterns** in information that enable it to complete tasks more efficiently and to a higher standard.

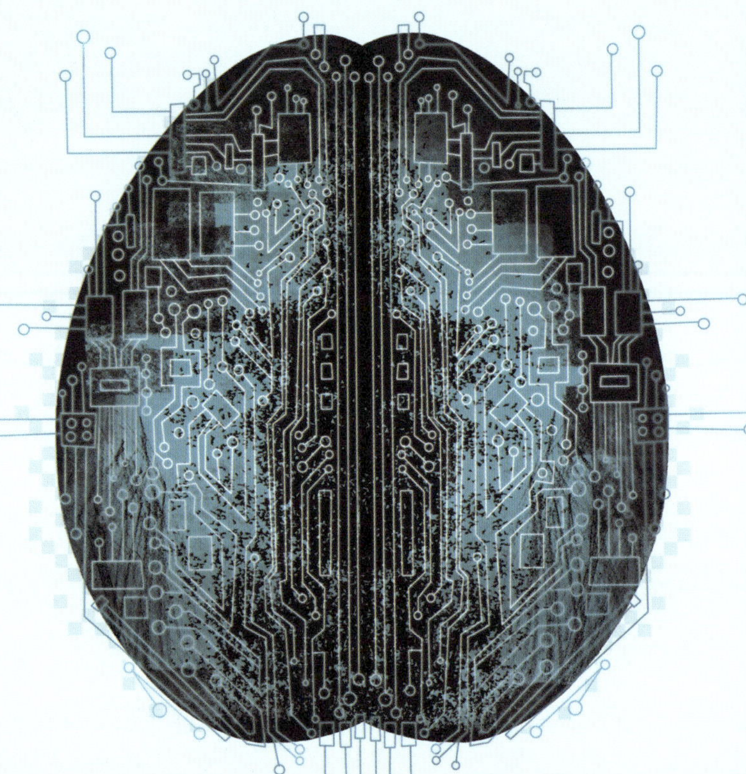

Founder of modern computing **Alan Turing** (1912-1954) read Ada's notes, almost a century after her time, as he explored the idea of "thinking" machines.

In Turing's 1950 paper "Computing Machinery and Intelligence," he referred to Ada's work: a computer cannot be said to have a mind, unless it could originate something entirely itself.

In the "Turing Test" a human tries to work out whether responses to their questions come from a human or a computer.

Scientists have since developed "Lovelace's Test," which is used to explore whether a computer is capable of original creations.

AN UNREADY WORLD
A Big Idea is Put on Pause

After the publication of her ground-breaking notes on Charles Babbage's new computing invention, the Analytical Engine, Ada wrote to Charles, offering her help in bringing the machine to life.

Though Charles was an excellent engineer and designer, he was terrible with public relations and making friends with the rich and powerful people who could help him.

Ada's long and detailed letter put forward her carefully thought-out plan.

Ada had a unique ability to understand and explain clearly the qualities of Babbage's complicated invention.

She also had excellent social skills and a network of wealthy friends.

Charles flatly refused her offer. Was he too protective of his idea, or too proud to accept the help of a woman?

In May 1851, the Great Exhibition opened in London.
It was a high point of Victorian science and technology.

Six million people visited
the amazing, purpose-built
glass pavilion in Hyde Park
called the Crystal Palace.

The newspapers were
full of reports of the
exhibits to be seen there:
exotic stuffed animals,
marble statues, and
even one of the world's
largest diamonds.

As her carriage arrived
outside the towering
glass palace, Ada saw
people from all over
the world gathered
to celebrate the
power of invention.

Queen Victoria's husband, Prince Albert,
had organized the exhibition to showcase Britain
as a leader in innovation. All manner of new
inventions were waiting inside.

Inside the exhibition, Ada saw the finest creations sent from all over the world.

Ada's husband was displaying decorative designs from the brickworks he owned. They won a medal!

Charles Babbage was among the crowds.
He was handing out a flyer about Mechanical
Notation—his system for describing and
understanding the workings of machines.

The thrill of creativity and development was in
the air everywhere. Surely the time was right for
the Analytical Engine?

The Great Exhibition was a triumph of modern invention at the time, and later became a symbol of the Victorian age.

But did any of the wonders on show have the potential that Ada predicted for Charles's design?

Regardless of its potential,
no further progress was
made, and the machine
remained unbuilt.

Still, the work that Charles
and Ada did complete was
an incredible achievement.

Though not celebrated
in her lifetime, Ada's
contribution to computing
had a lasting impact.

The Great EXHIBITION

Opening in May 1851, "The Great Exhibition of the Works of Industry of All Nations," was a celebration of Victorian invention in the United Kingdom.

It **showcased** over 100,000 objects from around the globe, from machinery and manufactured goods to fine treasures and exotic wonders.

The exhibition was the project of Queen Victoria's husband, Prince Albert.

The Koh-i-Noor—one of the world's largest **diamonds**—was a star attraction. Later, Queen Victoria wore it as a brooch.

Charles Dickens

Charlotte Brontë

Among the visitors were
authors Charles Dickens
and Charlotte Brontë and
scientist Charles Darwin,
who was working on his
groundbreaking theories
on evolution.

Charles Darwin

Charles Babbage's application to display plans
and models of his calculating engines at the Great
Exhibition were **rejected**, much to his annoyance.

QUESTIONING UNTIL THE END
The Digital Revolution

In 1852, Ada sat at her piano to pose for her portrait to be painted. The artist's father had painted Ada's father, Lord Byron, many years before. Now it was Ada's turn.

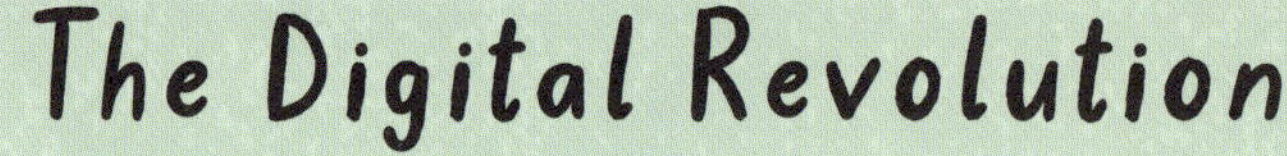

Ada, however, was feeling very unwell.

Though her scientific mind was at its strongest, she could feel her body was becoming weaker.

In a letter to a friend, Ada wrote that she was trying to find out what was wrong with her. She wanted to discover a system or method of understanding her health.

It was yet another example of her lifelong dedication to understanding the mysteries of the world.

Ada's doctors advised rest and time away at the seaside to help her back to health.

Soon though, it became clear she would not recover.

Her good friend, author Charles Dickens, visited her at her bedside and read scenes from his novels.

Ada's thoughts were never far from her father, Lord Byron, who had died when she was just eight years old.

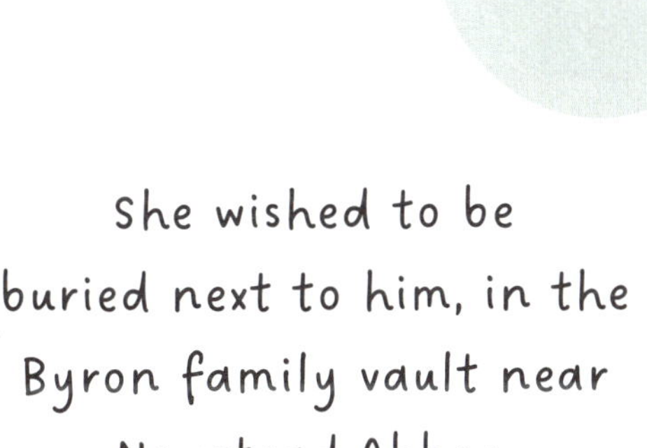

She wished to be buried next to him, in the Byron family vault near Newstead Abbey.

Ada died of cancer in November 1852, aged just 36 years old—the same age as her father when he died.

Ada did not live to see the the dawn of the computerized world we know today.

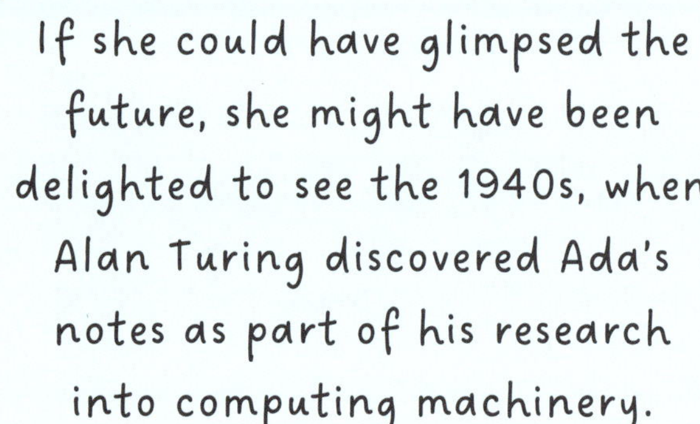

If she could have glimpsed the future, she might have been delighted to see the 1940s, when Alan Turing discovered Ada's notes as part of his research into computing machinery.

During the Second World War, Turing designed a code-breaking machine known as the Bombe.

It was used to de-code thousands of German military secrets.

The Bombe was operated by a female workforce— the Women's Royal Naval Service.

Turing's giant leaps in technology then paved the way for a digital revolution that would change the world.

Had Ada not been held back by society's expectations of her as a woman, the computer age may have begun much sooner.

Perhaps Ada and Charles would have lived happier
lives if they had known about the incredible legacy
they were leaving behind.

While neither of them ever touched a
fully-operational computer, their ideas
were instrumental in technologies that
have transformed the modern world.

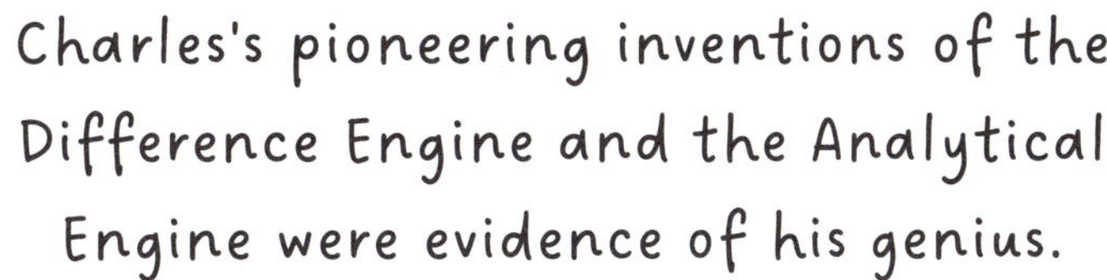

Charles's pioneering inventions of the Difference Engine and the Analytical Engine were evidence of his genius.

Ada's creativity and clarity of thought was the perfect match.

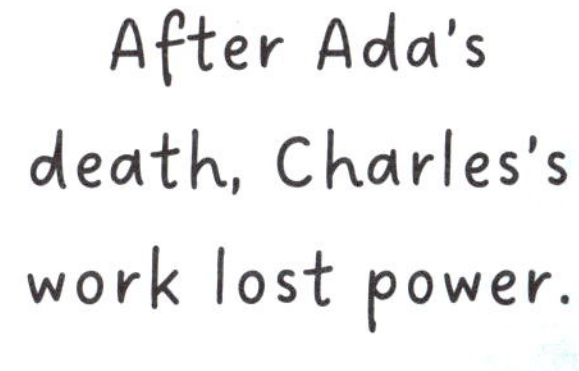

After Ada's death, Charles's work lost power.

Without Ada's clear explanations of the workings and potential of his machines, they remained only ideas.

A Remarkable LEGACY

Today, Ada is recognized as the very first complex computer programmer.

Ada realized that machines could perform certain tasks that previously could only be done by **humans**.

Computers use algorithms to choose the best chess moves.

She also saw that machines could further human **knowledge** and reduce error.

A.I. analyzes medical scans to detect disease early.

Each year, the second Tuesday in
October is "**Lovelace Day**."
It's a celebration of women in
science all over the world.

Ada had a sense of
her own genius in her
lifetime, but it took
years going by to reveal
her full legacy.

As Ada said herself: "That brain of mine is something
more than merely **mortal**; as time will show."

ADA LOVELACE'S JOURNEY TO BECOMING THE FIRST COMPUTER PROGRAMMER

1815

December 10

Augusta Ada Byron is **born** in England.

1817

Ada's father, **Lord Byron,** leaves for Italy.

1824

Ada's father dies while involved in the **Greek War of Independence.**

1835

Ada marries **William King.**

1834

Ada and her mother tour the **textile factories** of northern England.

1836

Ada gives birth to her **first son,** Byron.

1837

Ada gives birth to her **daughter,** Anne Isabella.

1838

William King is named Earl of Lovelace, Ada becomes **Countess of Lovelace.**

1979

A **computer programming language** developed by the US Department of Defense is named "Ada" in her honor.

1852

November 27

Ada becomes very ill and **dies** of cancer, aged 36.

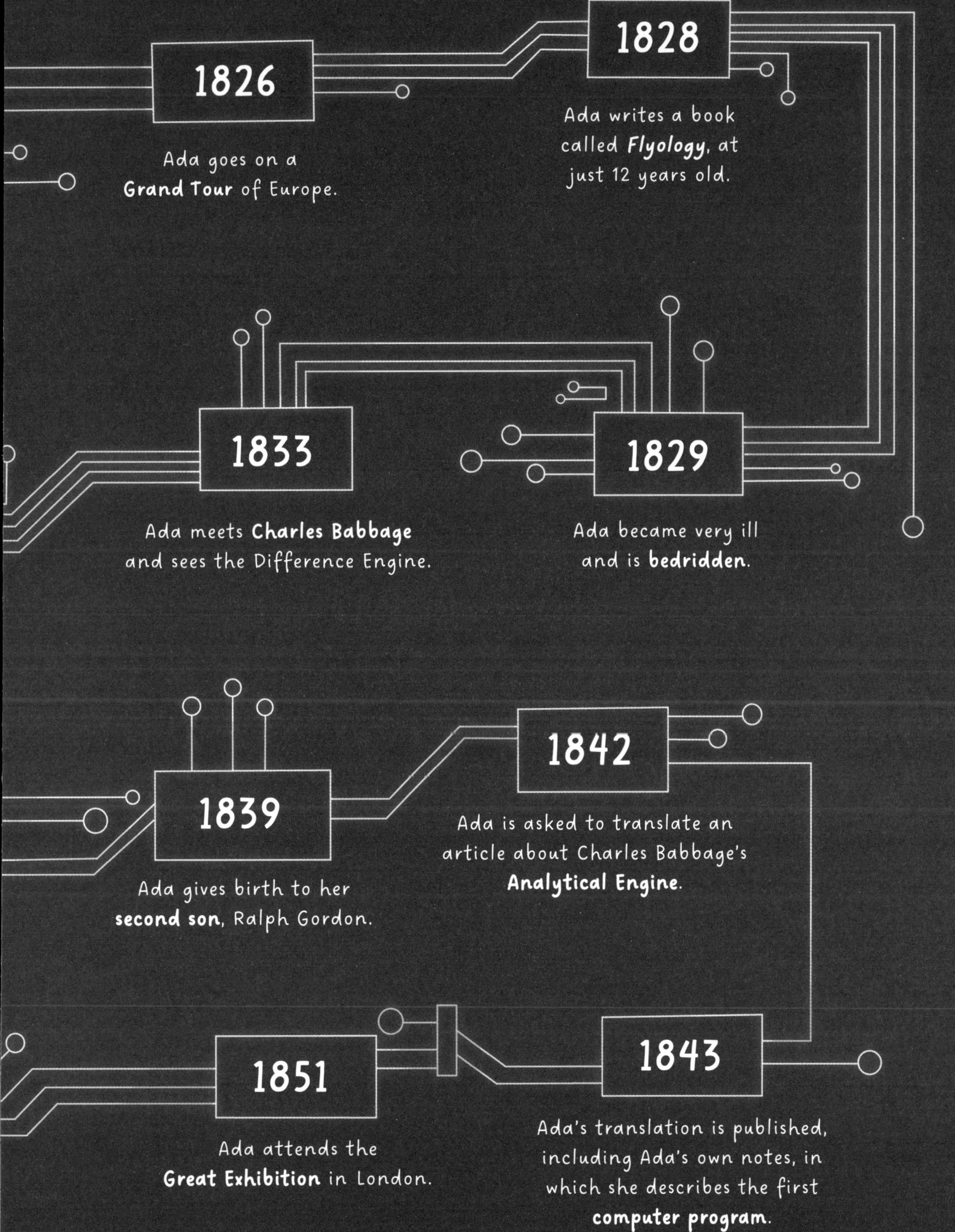

1826
Ada goes on a Grand Tour of Europe.

1828
Ada writes a book called **Flyology**, at just 12 years old.

1833
Ada meets **Charles Babbage** and sees the Difference Engine.

1829
Ada became very ill and is **bedridden**.

1839
Ada gives birth to her **second son**, Ralph Gordon.

1842
Ada is asked to translate an article about Charles Babbage's **Analytical Engine**.

1851
Ada attends the **Great Exhibition** in London.

1843
Ada's translation is published, including Ada's own notes, in which she describes the first **computer program**.